Points of VIEW

·P·O·L·L·U·T·I·O·N·

Barbara James

Wayland

Points of View

Abortion
Advertising
Alcohol
Animal Rights
Censorship
Crime and Punishment
Divorce
Drugs
Medical Ethics
Northern Ireland
Nuclear Weapons
Pollution
Racism
Sex and Sexuality
Smoking
Terrorism

Series Editor: William Wharfe
Editors: Sue Hadden/Paul Bennett
Designer: David Armitage

First published in 1991 by
Wayland (Publishers) Limited
61 Western Road, Hove
East Sussex BN3 1JD, England

British Library Cataloguing in Publication Data
James, Barbara
Pollution. – (Points of view)
I. Title II. Series
333.7

ISBN 0-7502-0198-3

Phototypeset by Direct Image Photosetting Ltd, Hove, East Sussex, England
Printed in Italy by G. Canale & C.S.p.A, Turin
Bound in France by A.G.M.

Acknowledgements

The publishers gratefully acknowledge permission from the following to reproduce copyright material: *Ambio*, article January, 1990; The American Assembly, *Preserving the Global Economy: The Challenge for Shared Leadership*, ed Jessica Tuchman Mathews; Ark Campaigns, *The DIY Guide to Combating Global Warming*, Jeremy Rifkin and Mick Kelly, 1990; BBC Enterprises Ltd, *What on Earth are We Doing?* Jonathon Porritt, 1990; British Gas plc 1) *Green Economics*, Dr Anil Marklandya, British Gas Key Environmental Issues No. 8; 2) *The New Environmental Policy*, David Pearce, British Gas Key Environmental Issues Series No. 1; CBI, *Clean up – It's Good Business*, 1986; Century, *How to be Green*, John Button, 1989; Collins 1) *Turning the Tide*, David Bellamy and Brendan Quayle, 1986; 2) *Life on Earth*, David Attenborough, 1979; Commission of the European Communities, *Environmental Policy in the European Community*, May 1990; *Country Life*, Geoffrey Lean, November 1990; *The Economist*, 20 Oct, 1990; Environment Canada, *Canada's Environment: An Overview*, 1986; Erik Eckholm, *Down to Earth*, 1982; *The Field*, article by Frances S Wilson, December 1990; Friends of the Earth, *Earth Matters*, 1) Koy Thompson, No.5; 2) Autumn 1989; Green Drum, *A Green Look at the Economy*, Jonathan Gray, Summer 1989; Greenpeace, *Campaign Report*, November 1990; the *Guardian*, 18 May 1988, 22 Aug 1989; HMSO: 1) *This Common Inheritance*, White Paper, 1990; 2) *Protecting Your Environment*, DOE, 1989; Hyde Park Publications, *Environment Now*, (no.12) articles by Jonathon Porritt, Sara Parkin, and Anne-Maria Brennan; IIED/Earthscan Publications, *Our Common Future: A Reader's Guide*, Don Hinrichsen, 1987; the *Independent*: Patricia Clough, 23 October 1990; 2) Bill McKibben, 27 January 1990; 3) Nicholas Schoon, 17 April 1990; Green Paint, *The Stolen Future*, Patrick Rivers, 1988; 4) James Lovelock, 23 September 1989; IUCN: 1) *IUCN Bulletin Vol. 20, 1989; 2) World Conservation Strategy*; Longman Group UK, *Longman Chronicle of the 20th Century*, ed. D Mercer, 1988; National Society for Clean Air, *NSCA Reference Book*, 1989; W.W. Norton & Company, *The Twenty-Ninth Day*, Lester Brown, 1978; Optima 1) *Friends of the Earth Handbook*, 1987; 2) *Green Manifesto*, Irvine and Ponton, 1988; Oxford University Press, *Our Common Future*, World Commission on the Environment and Development, 1987; Penguin, *Poverty and the Planet*, Ben Jackson, 1990; Pluto Press, *The State of the Earth*, ed Joni Seager, 1990; *The New Road*, the Bulletin of the WWF Network on Conservation and Religion; Scope, *The Globe Greenhouse*, John Gribbin, 1989; *The Sunday Times Magzine*, 26 February 1989; UNEP, *Our Planet*: 1) Share Care, Vol.2 no.2, 1990; 2) editorial Vol. 1, no.4, 1989; 3) Vol.1 no.2/3, 1989; Usborne, *Dictionary of Science*, C. Stockley, C. Oxlade and J. Wertheim, 1988; Victor Gollancz, *The Green Capitalists*, John Elkington, 1987; Virago Press, *The Young Person's Guide to Saving the Planet*, Debbie Silver and Bernadette Vallely, 1990; WHO, *World Health*, January-February, 1990; Writers and Readers/ Unwin Paperbacks, *Capitalism for Beginners*, Lekachman, R. and Loon, B. van, 1986; WWF International, *WWF News*, No.66, 1990.

Contents

1 Introduction: pollution in the world

Kuwaiti oil wells sabotaged during the Gulf war burn out of control. The thick, black smoke from over 500 burning wells has caused pollution on a massive scale affecting the health of Kuwaiti people and the ecology of the region.

> In the history of pollution no name is etched more corrosively than that of Bhopal. In December 1984 this central Indian town was the site of a major chemical accident involving a pesticide plant belong to [the chemical company] Union Carbide. In the middle of the night an invisible cloud of toxic gas leaked out from its tank and enveloped a nearby slum housing 200,000 people . . . the total figure [of deaths] could be as high as 10,000 . . . (*The Sunday Times Magazine*, 26 February 1989.)

The pollution disaster at Bhopal was widely reported in the world's media but, all over the world, there have been many more pollution incidents. Pollution is not always dramatic – it can work away slowly and quietly for years before the scale of the problem is realized and understood. Pollution is a serious, and potentially deadly, environmental issue. It affects every one of us.

• What is pollution?

Pollution has been defined as:

> . . . the release into the land, atmosphere, rivers and oceans of undesirable substances which upset the natural processes of the Earth. These substances are known as pollutants. (Stockley, Oxlade and Wertheim, *Dictionary of Science*.)

Very often these pollutants are waste products from industrial, agricultural or domestic processes. The Earth has natural processes which are very good at breaking down and recycling natural wastes.

> . . . the absorption of waste is an important natural function of the Earth's ecosystem. Indeed, in the complex web of plant and animal life, one organism's waste is another's sustenance [food]. But when waste is excessive, it becomes pollution. (Lester Brown, *The Twenty-Ninth Day*.)

Waste has become excessive in some environments and the natural recycling systems are overloaded. Humans are usually the cause.

> The Earth is the birthplace of our species and, so far as we know, our only home. When our numbers were small and our technology feeble, we were powerless to influence the environment of our world. But today, suddenly, almost without anyone noticing, our numbers have become immense and our technology has achieved vast, even awesome, powers. Intentionally or inadvertently, we are now able to make devastating changes in the global environment . . . (Global Forum of Spiritual and Parliamentary Leaders, Moscow Forum, January 1990.)

The explosion at the Chernobyl power plant in 1986 was the world's worst nuclear accident. Over 100,000 people were evacuated from their homes and the radioactive fall-out contaminated large areas of Europe. The accident changed public attitudes on nuclear power, and since then at least five countries have decided not to build more nuclear power plants.

• Types of pollution

Pollution of the land, sea and air occurs as a result of many human activities. Waste chemicals from industry are often poured into rivers or discharged as gases into the air. Waste from coal mines and steel works forms ugly heaps on the land. Agriculture also pollutes the environment, with pesticide sprays drifting in the air and fertilizers affecting plant life in rivers and ponds. A lot of pollution comes from our homes – sewage often ends up in the sea, while the mass of domestic rubbish has to be disposed of at dumps or landfill sites. Accidental pollution can be the most dangerous – think of the damage caused by oil slicks or an accident at a nuclear power station.

It is often hard to measure pollution but some estimates are most alarming.

> Italy's River Po discharges 230 tons of arsenic, 18,000 tons of phosphates and 135,000 tons of nitrates into the Adriatic Sea each year. In the past two years the consequences have been disastrous for Adriatic seaside resorts. (*World Health*, January/February 1990.)

Pollution is also caused by natural phenomena, such as volcanoes. They pour huge amounts of chemicals into the atmosphere, including the gas, sulphur dioxide, a major cause of acid rain.

> Fifteen hundred people died today in the Cameroons when a deadly mixture of carbon dioxide and hydrogen gas escaped from a volcanic lake near the town of Wum. (*Longman Chronicle of the 20th Century*, 1988.)

A grey seal dies from the effects of oil pollution at sea. Oil pollution can be accidental, such as a leakage from a tanker collision, or deliberate, such as when tankers wash out their holds at sea.

The end product of the consumer society is waste. Pollution is usually caused by waste from industry, agriculture or homes. There is too much waste to be broken down naturally, so how can we dispose of so much waste safely?

• What are the effects of pollution?

> Pollution is more than a mere nuisance . . . It can ruin forests, crops and fisheries . . . freshwater lakes and streams; destroy whole species of plants and animals; impair [harm] human health; break up the ozone layer; . . . [harm the] atmosphere; and even damage clothing, buildings and statues. (Lester Brown, *The Twenty-Ninth Day*.)

The effects of pollution on the environment are often all too obvious, but what about the risks facing people? The casualties of the Bhopal accident prove how devastating chemical pollution can be. But do we understand the effects of 'everyday' pollution on millions of people?

> In the Soviet Union, government figures show that 50 million people breathe air that contains ten times the pollutants allowed by [their] air pollution laws. (Shane Cave, *Our Planet*, Vol. 2.)

Pollution is ugly – and it can be deadly. No one wants it and no one thinks it is a good thing. So why does it still happen? Why is it a major world problem? The answers to these questions are complex and lie deep within our history and traditions. In this book we shall look at some of these questions and answers. We shall examine the points of view of the polluters, the environmentalists, the politicians and the public.

Space pollution

Pollution has now spread outside the planet, mostly in a region between 160 and 1,760 km above the Earth. There is no internationally accepted body that manages outer space.

'The growing pollution of orbital space with debris [rubbish] from space weapons tests, spent rocket parts, fuel tanks and dead satellites could pose an "unacceptable threat" to life in space within a decade.' (Don Hinrichsen, *A Reader's Guide to Our Common Future*.)

2

The moral issues

Morals are concerned with human behaviour. Is an action right or wrong, good or bad? Although societies may have accepted standards of conduct, many people have their own moral codes based on their background and beliefs. Many environmental issues question our moral values.

• Crimes against Creation

Most pollution in the world has been the result of human activity.

> We are now threatened by self-inflicted, swiftly moving environmental alterations about whose long-term biological and ecological consequences we are still painfully ignorant – depletion of the protective ozone layer; a global warming unprecedented [never known] in the last 150 millenia; the obliteration of an acre of forest every second . . . We are close to committing – many would argue we are already committing – what in religious language are sometimes called Crimes against Creation. (*Preserving and Cherishing the Earth*, Global Forum of Spiritual and Parliamentary Leaders, January 1990.)

Pollution can kill humans and other species and damage the planet. What right do humans have to destroy or abuse other forms of life on the Earth?

The beauty of nature is legally protected here in the Denali National Park, Alaska, but many areas of the world are not so lucky. How people feel about themselves and the world has an affect on the way they treat the environment.

> . . . the most profound moral question behind the environmental crisis concerns the lack of respect for life. (His Holiness Pope John Paul II, New Year message 1989.)

Why is there this lack of respect for other forms of life? Why do some humans behave in this way? The answers to questions like these are complex, but there are some clues to human behaviour in our religions and belief systems. Most Western countries follow the Jewish-Christian tradition. The following extract from the Bible concerns Adam and Eve.

> And God blessed them, and God said to them, 'be fruitful and multiply, and fill the Earth and subdue it; and have dominion over the fish of the sea and over the birds of the air and every living thing that moves upon the Earth.' (Genesis Chapter 1, verse 28.)

This verse has often been interpreted by humans as a God-given right for them to use the Earth as they please. But do humans have a right to do this?

> God's command that Adam should conquer the world has been misunderstood for centuries. God did not really mean conquer and destroy, as some theologians have interpreted in the past, but take care of creation, govern rather than destroy. (Dr Arnd Wünschmann, Director of Conservation, World Wide Fund for Nature, Germany.)

Elsewhere in the Bible, the words of Jesus suggest great respect for the beauty of nature.

> See how the lilies of the field grow. They do not labour or spin. Yet I tell you that not even Solomon in all his splendour was dressed like one of these. (Matthew, Chapter 6, verses 28-9.)

But what about other religions? The Buddhist religion advocates living in harmony with the environment.

> We regard our survival as an undeniable right. As co-inhabitants of this planet, other species too have this right for survival. (*The Buddhist Declaration on Nature.*)

Islam sees humans as stewards of God's creation:

> For the Muslim, mankind's role on Earth is that of a 'khalifa', vice-regent or trustee of God. We are God's stewards and agents on Earth. We are not masters of this Earth, it does not belong to us to do what we wish. It belongs to God and He has entrusted us with its safekeeping. (*The Islamic Declaration on Nature.*)

Pope John Paul II has reminded the people of the world that it is threatened by a lack of respect for nature. He says, 'If man is not at peace with God, the Earth itself is not at peace.' Do you agree with this statement?

What sort of world will your children and grandchildren inherit? What is their future? How do you think they will view the past generations of people who treated the Earth so carelessly?

Whether or not we have a moral responsibility towards other forms of life on Earth, many people feel that, as part of Creation, we have a duty to safeguard the lives of future generations. Pollution, and the environmental damage it causes, lowers the quality of our own lives and those of our children, grandchildren and future generations. The Tibetan spiritual leader, the Dalai Lama, reminds us of our responsibility:

> We are sacrificing the fate of future generations to our own short-term interests. This is an extremely selfish and morally wrong attitude. (His Holiness the Dalai Lama.)

> No generation has a freehold on this Earth. All we have is a life tenancy – with a full repairing lease. (Margaret Thatcher, British Prime Minister 1979/90.)

● Pollution across frontiers

Another issue in the pollution debate is transboundary pollution. Environmental damage does not stop at national boundaries; pollution from one country can affect another. Sometimes this is accidental, but it can also be deliberate.

> The global environment is a commons, shared by all countries. Increasingly, countries also share in its deterioration. Most pollution and degradation [worsening] of the environment cannot be contained within national borders . . . In most cases, this share-out of environmental problems is inevitable. Occasionally, though, governments and private industries take active measures to 'ship' their pollution abroad. They may build smokestacks tall enough to disperse airborne pollutants, for example, or export unwanted and hazardous waste to poorer countries. (Joni Seager, *The State of the Earth.*)

Scandinavia suffers acid rain damage caused mainly by sulphur dioxide pollution blown by the winds from factories and power stations in other European countries. The acid rain has damaged Swedish and Norwegian lakes and forests, resulting in the death of fish, birds and other wildlife. It has polluted some drinking water supplies and corroded historic buildings. Acid rain can also ruin the livelihood of farmers and foresters. One Swede describes the damage:

> My family has lived on this farm for over 100 years. It covers about 29 hectares, 23 of which are forest. Every year a contractor comes in to fell some of the trees. The older ones are over 100 years old and fetch a high price, as they can be used for timber. However, many of these old trees are now damaged by the acid rain . . . The younger trees will never reach this age. (Ingemar Zachrisson quoted in *Acid Rain* by John Baines.)

Do you think the polluting countries have a moral responsibility to make every effort to reduce their pollution? Or should it be left to the country that is affected to deal with the problem?

Acid rain has contributed to the destruction of huge areas of forest in Europe and North America, such as this region on the German-Czechoslovakian border. Pollution cannot be kept within national boundaries, so international measures are needed to deal with it.

• Trading in pollution

One controversial environmental issue is the deliberate dumping of polluted waste by developed countries in poorer, developing countries and, more recently, in Eastern Europe. The waste may be highly toxic (poisonous) and nobody wants it in their 'back yard'. Often poorer countries have less strict pollution control regulations, or are desperate for foreign currency to pay off the high interest rates on their debts. Developed countries are prepared to pay a lot to have their waste disposed of.

> Toxic waste from rich Western Europe, barred from many Third World countries after a succession of scandals, is now heading east, to impoverished and highly polluted Poland . . . Since Poland opened its borders to business with the West, 22 million tons of poisonous waste are known to have been offered. Of these, at least 47,000 tons have arrived, compounding the country's severe environmental problems. (Patricia Clough, *The Independent*, 23 October, 1990.)

Is this a justifiable trade or are the richer countries exploiting poorer ones, such as Poland? How do we deal with poisonous waste when nobody wants it?

• Healing the wounds

> We have sinned against the environment and the world, and our repentance lies in dedicating ourselves to preserving the environment. (Simon Barrington Ward, Bishop of Coventry.)

The Greenpeace vessel Sirius *monitors the* Karin B, *a freighter that was hired to carry toxic waste from Italy to be dumped in Nigeria. When the Nigerians refused entry to the waste, the* Karin B *returned to Europe only to be refused entry to many ports. No one would accept her deadly cargo and she eventually returned to Italy.*

What are the solutions to these moral issues surrounding human use and abuse of the environment? Religious leaders agree that a moral revolution is needed.

> The ecology crisis is proof of the urgent moral need for a new solidarity [togetherness], especially between developing nations and highly developed, industrialized countries. Today the ecological problem has assumed such dimensions that it demands a universal response. (His Holiness Pope John Paul II, New Year message 1989.)

> We should begin to develop . . . what I call a sense of universal responsibility – a deep respect for every living being who lives on this one small planet and calls it home. (His Holiness the Dalai Lama.)

This may seem impossible but maybe it is a question of rediscovering what we already know.

> So the central challenge is – how do we develop a society whose roots are in conservation – in harmony with the Earth? Many indigenous [original] peoples can provide traditional wisdom and traditional methods of living in harmony with the environment . . . we need to incorporate the best in their traditional wisdom. (M S Swaminathan, *IUCN Bulletin* Vol. 20.)

Now that more people understand the problems facing our planet, positive changes are happening all over the world. There is now a worldwide network of religious and conservation organizations. The world's major faiths have made statements on the environment and human responsibility towards it. But it is also up to individuals.

> You damage the Earth just by living on it. You burn fossil fuels – petrol, oil, coal – and huge amounts are burnt by those who supply you with goods and services. You create waste, which has to be buried, burnt or discharged into the sea. You accept the profits of investments which are trading on Third World poverty and putting further strain on already over-stretched resources. You buy goods from farms and factories whose ill effects from chemical waste range all the way from dead fish to dead people. You are a polluter. (*The Sunday Times Magazine*, 26 February, 1989.)

How to care for the environment is a moral decision for all of us. It can affect how we act, how we live, what we buy, who we vote for, how we think and feel about our world.

The Dalai Lama is the spiritual leader of Tibetan Buddhists. He calls on every one of us to consider our relationship with the Earth.

1 To whom do we have a moral responsibility:
ourselves?
our families?
our country?
our species?
other species?
the planet?

2 Do you think humans are more important than other species?
Give reasons for your answer.

3

Economics and the environment

Economic growth in Western countries has resulted in a huge increase in the ownership and use of cars, but what are the effects of so many cars on the environment? Air pollution, traffic jams, loss of countryside to roads and a waste of a valuable resource – energy – are the price we pay for freedom of movement.

Most people living in Western or developed countries have comfortable lives. They have homes, jobs, cars, televisions, well-stocked kitchens, education and health services. Since the 1950s the standard of living has increased rapidly:

> For twenty years after the Second World War, capitalism enjoyed a prolonged and quite unprecedented boom. The world system had never grown so fast and for so long – twice as fast between 1950 and 1964 as between 1913 and 1950. (*Capitalism for Beginners.*)

This increasing level of wealth has been achieved by economic growth based on cheap sources of energy and raw materials. Economic growth was widely seen as a good thing – it provided jobs and material security for the population, prompting British Prime Minister Harold Macmillan to claim in 1957 'You've never had it so good.'

However, the effect of rapid economic growth on the environment was not taken into consideration.

> In the early sixties . . . the economics profession devoted a lot of attention to developing theories of growth. These theories were meant to show how an economy could achieve unlimited stable growth for the indefinite future . . . No account was taken of the limited capacity of the environment to sustain such growth. (Dr Anil Markandya, *Green Economics*, British Gas Key Environmental Issues Series No. 8.)

What effect does economic growth have on the environment? As the quotation below explains, the main effect is on the Earth's natural resources, which are fragile and finite.

> Serious thinkers are beginning to realize that man's interest in economic growth may conflict with his interest in maintaining a high-quality environment. Economic growth . . . requires large amounts of land, forests, water and other natural resources. As these resources are consumed to produce houses, automobiles and televisions, they no longer exist as amenities. The time has come when serious choices must be made between growth and [environmental] quality. (Barkley & Seckler, *Economic Growth and Decay.*)

Some people still support the growth theory, but in a world of limited resources, is this possible? Many environmentalists think not.

> [The pursuit of economic growth has caused] massive environmental havoc . . . industrial pollution, wasteful resource depletion, biological destruction and climatic disruption . . . [In view of this,] many people, and particularly Greens, are claiming that investment in such growth has long since reached saturation point. (Jonathan Gray, *A Green Looks at the Economy.*)

But who is going to make choices between growth and environmental quality? Few people want to lower their standard of living and any moves by governments to do this would be very unpopular. In 1989 the Dutch Government was brought down by a row over how to finance a National Environment Plan, which would have imposed taxes to pay for pollution.

Many people in developed countries live comfortable lives with many luxuries. How many people would lower their standard of living to reduce human impact on the environment, or to help developing countries? Should people be forced to consume less by law?

> **Who consumes what?**
>
> In a year, the average person in a developed country is likely to consume (directly or indirectly):
>
> ★ more than 120 kg of paper
> ★ over 450 kg of steel
> ★ 26 kg of other metals
> ★ energy equivalent to 6 tonnes of coal
>
> In a developing country, the average person is likely to consume:
>
> ★ 8 kg of paper
> ★ 43 kg of steel
> ★ 2 kg of other metals
> ★ energy equivalent of 0.5 tonnes of coal
>
> Source: *The Green Consumer Guide*

> It is very difficult for any government to abandon economic growth. It is asking almost the impossible. Yet I don't quite see how we are going to go on having economic growth when there is no growth left in the world's ecology. At some stage one is going to destroy the other. (His Royal Highness, Prince Philip, quoted in *The Guardian*, 18 May, 1988.)

We have seen how the richer, industrialized countries can damage the environment. But the situation is complex, because the opposite of economic prosperity – poverty – can also damage the environment.

> . . . the daily survival decisions of the poor and hungry disrupt their own life-support systems, impair ecological processes and destroy genetic and other renewable resources just as surely as do too many of the development decisions of the rich and powerful in government and industry. (*World Conservation Strategy*.)

There are estimated to be one billion people, nearly a fifth of the world's population, living in 'absolute poverty'. Absolute poverty means disease, malnutrition, illiteracy, a high infant death rate and low life expectancy. How can these people even consider the environment when their own survival is uncertain? Can economic growth change their situation?

> Without economic growth we cannot hope to defeat poverty, the greatest environmental wrecker. (Dr Mostafa Tolba, Executive Director of the United Nations Environment Programme (UNEP), in a speech to UNEP UK Committee, 1988.)

Economists and environmentalists have come to understand that their subjects are linked and this has generated a new area of discussion.

This clothing factory in the Philippines employs about 1,000 people, mostly women. Many industries and resources in developing countries are used to supply the insatiable demand for goods in Western countries.

Education is the key. This is a village school in Pakistan, but people all over the world need to learn about the environment and to understand how to live sustainably on the planet.

> The issue in the new environmental debate, then, is not whether we grow or not. It is **how** we grow. Environmental economics tells us there are ways of giving people what they want . . . and maintaining environmental quality. (David Pearce, *The New Environmental Policy*, British Gas Key Environmental Issues Series No. 1.)

● Sustainable development

So what are these ways of having economic growth in a way that is environmentally acceptable? In the 1970s the concept of sustainable development began to emerge. Sustainable development is defined as:

> Development that meets the needs of the present without compromising [reducing] the ability of future generations to meet their own needs. (World Commission on Environment and Development, *Our Common Future*.)

Many governments and industries are attempting to put this concept into practice. In 1990 the British Government published its environmental strategy. It says:

> We have a moral duty to look after our planet and hand it on in good order to future generations. That does not mean trying to halt economic growth. We need growth to give us the means to live better and healthier lives. Technological development can help clean up our surroundings. But growth has to respect the environment. And it must be soundly based so that it can last. That is what is meant by sustainable development. (*This Common Inheritance*. A summary of the government report (White Paper) on the environment.)

An economic benefit to one person may be a loss to someone else. Read the quote from Ingemar Zachrisson in chapter two again. Should he be paid compensation for the loss of income on his farm? If so, who should pay for it – the Swedish Government, the governments of the polluting countries, the industries that cause the pollution or every one of us through taxation or higher prices?

4 Who pays the price?

> The industrialized countries bear the greatest responsibility for the pollution of the environment. (José Sarney, former President of Brazil.)

It is certainly true that the developed (industrialized) countries are responsible for huge amounts of pollution. They use the most resources and produce the most waste. However, poverty as well as wealth can lead to pollution.

> Environmental stress . . . [arises from] the growing demand on scarce resources and the pollution generated by the rising living standards of the relatively affluent [the more wealthy nations]. But poverty itself pollutes the environment, creating environmental stress in a different way. Those who are poor and hungry will often destroy their immediate environment in order to survive . . . (World Commission on Environment and Development, *Our Common Future.*)

● The quest for development

In some developing countries populations are rising rapidly. Huge numbers of people need food, fuel, shelter and work. This puts pressure on natural resources and care for the environment comes a poor second to survival. People living in poor countries have as much right as anyone else to improve their standard of living, in just the way developed countries have already done.

People's need for food and to make a living can degrade the environment. Here in Brazil, huge areas of tropical forest have been felled and burnt to provide agricultural land. Many developed countries, such as Britain, deforested their own land for agriculture centuries ago. Do we have a right to tell the Brazilians what to do?

> The quest for prosperity is a universal human impulse. The poor want to be rescued from their wretchedness. The rich want at least to remain rich, if not get richer. Countries, like individuals, nurture the same impulses. (Barber Conable, President of the World Bank.)

Developing countries want to improve their standard of living. To do this, their economies need to grow. They need to develop industry and build modern factories. Iron and steel production, brick-making and the manufacture of chemicals are all important industries for the developing countries. But they also cause much pollution.

To finance much of this development, poorer countries need 'hard currency' (Western currencies) to buy Western technology and goods. Most developing countries borrow this money from banks in developed countries but the loans have to be paid back with interest. Many countries find that they are unable to pay back the loans and, as interest rates go up, they face mounting debt as this article illustrates:

> Developing countries owe rich countries over $1,300 billion . . . huge debt repayments mean that there is actually a net outflow from Third World to richer countries ($40 billion in 1988). Worse still, this outflow does not even cover the interest due, so [the developing countries] borrow, piling debt upon debt. (Koy Thomson, *Earth Matters* No. 5, 1989).

In order to raise the money to pay the debt, poor countries may have to sacrifice environmental concerns for a quick return on investment.

> Rising poverty and desperate attempts of Third World countries to earn hard currency carry not only a heavy human and economic cost, but also an environmental one. Many [developing] nations rely on their natural resources – whether timber or minerals like copper – to raise foreign exchange from trade [and so pay off their debts to Western countries]. A Philippine grass-roots network, the Freedom from Debt Coalition, says that debt is being paid for by: '. . . tearing down more of our forests for lumber [timber] exports, destroying more of our coral reefs for fish exports, and depleting more of our soil through heavy pesticides and fertilizers used for cash crops.' (Ben Jackson, *Poverty and the Planet.*)

International debt causes many countries to ignore the environment in an effort to earn foreign currency. Here, in Indonesia, ebony trees from tropical forests are felled so the wood can be sold abroad in exchange for hard currencies.

So, how can these countries develop? The world population was 5 billion in 1990; it will be 6 billion by 2000 and 8 billion by 2025. Most of the people will be born in developing countries.

> The countries which will house those teeming millions will face – are already facing – enormous pressures for industrialization. Can they possibly follow our example when, as we have seen, our activities have already damaged the world? (A B Cleaver, Chief Executive of IBM UK.)

The price of industrial development is a polluted world, as we have seen. But industrial development offers an attractive, easy lifestyle.

> The gleaming bathroom and kitchen, the well-stocked refrigerator, his-and-her cars in the drive: for half a century that stereotyped model of well-being has been flashed across cinema screens from Tibet to Timbuktu. It is a dream . . . Disillusion is at hand: there will be 6 billion people by 1998 and they cannot live like that. (Walter Schwarz, *The Guardian*, 22 August, 1989.)

Above *Population pressure in Hong Kong means that housing has to be high-rise to accommodate everyone. The world population is set to increase by 50 per cent in the next thirty years. Living space may be found for them, but what about other resources, such as food, fuel and clean water?*

If the Earth cannot support everyone having the same standard of living as Western countries, who makes the decisions about how to share the resources out? In this quote, Jonathon Porritt, former Director of Friends of the Earth in Britain, outlines the views of many developing countries. He writes as though he were a spokesperson for them:

> You in the North [developed countries] got rich by doing what we now aspire to do: using your land, raw materials and energy sources to fuel a process of economic growth and industrial expansion. In the process you have caused untold ecological damage in your own countries and in ours. You now expect us to hold back from a pattern of development that has served you so well because you fear that in the process we will exacerbate [add to] the pollution you have already caused and accelerate the process of ecological collapse you first set in train. And you apparently expect us to do this out of the goodness of our hearts when millions of our people are dying of malnutrition or preventable diseases, or living in grinding poverty. Well, you can get stuffed – unless you are prepared to pay up. (Jonathon Porritt, *What on Earth Are We Doing?*)

Fair shares for all?
33 per cent of the world's population uses 85 per cent of the world's resources. The remaining 67 per cent uses only 15 per cent of the world's resources.

The Erickson power plant in Michigan, USA. Economic growth and industrialization require huge amounts of energy in the form of electricity, coal, gas and oil. All energy production has an effect on the environment.

Should developed countries pay up or at least wipe out the debt? There are no easy solutions to the development dilemma but there are some things that can be done to stop problems getting worse. Planning for sustainable development, as we have seen in chapter three, is vital to both developing and developed countries. Another way forward is through the transfer of technology from rich to poor countries, so that less polluting and more 'environment-friendly' processes are available to all.

> Technology development has improved the environment in the industrial parts of the world . . . The future for these [developing] countries does not look very bright, unless international society takes it upon itself to [share] production technology and resources. This is politically difficult indeed. (Juul Bjerke, International Confederation of Trade Unions.)

Below *Two Masai look at a geothermal power unit in the Kenyan Rift Valley. Developing countries need energy if their economies are to grow. Kenya is fortunate in being able to use geothermal energy because it is less polluting than fossil fuel.*

A cart delivers clean water in Calcutta. In many developing countries, millions of people suffer from diseases transmitted through dirty water. A clean water supply and proper sanitation for all people in the world could be easily achieved with international co-operation. The United Nations estimates that it would cost US$25 million a day – only 10 per cent of what is spent on cigarettes each day.

• And justice for all?

At the heart of the environment and development dilemma is the need for a fairer distribution of world resources. This can only come about by international goodwill and co-operation.

> It is increasingly clear that countries must act together to deal with environmental issues; that is an obligation born of our common history and destiny. But for this to be achieved, nothing less than a new world order is needed, a fundamental shift in national perceptions towards global co-operation. (Dr Mostafa Tolba, editorial in *Our Planet*, Vol. 1, No. 4, 1989.)

Development data

Country	Population 1989 (millions)	Population growth rate (percentage increase 1985–90)	No. of cars 1986–87 (thousands)
Australia	16.5	1.25	7,073
Bangladesh	112.3	2.61	45
Brazil	147.4	2.07	10,827
China	1,110.8	1.18	995
Ethiopia	48.7	2.79	41
Kenya	24.4	4.20	126
Saudi Arabia	13.5	3.84	2,245
UK	56.2	0.02	20,097
USA	246.3	0.86	137,324
USSR	289.3	0.93	11,800

1 Do you think a Western developed country (such as Britain or the USA) offers the only satisfying way of life?

2 Do you think you have a right to increase your standard of living? Do people in poor countries have the same rights as those in rich countries?

5

The greening of politics

> When you've spent half your political life dealing with humdrum issues like the environment . . . it's exciting to have a real crisis on your hands. (Margaret Thatcher, 14 May, 1982.)

This quote by the former British Prime Minister, Margaret Thatcher, was spoken during the Falklands Islands war in 1982. Things have changed a lot since then. Today the environment is a hot political issue with parties and politicians, both national and local, keen to be green. How has this change come about and are political parties as environmentally caring as they say, or are they out to win votes?

Greenpeace raises the profile of environmental issues by non-violent direct action. Here they are trying to stop the dumping of flyash waste from British power stations in the North Sea. Action like this ensures Greenpeace publicity which, in turn, affects public opinion.

Right *Public demonstrations are another method of environmental protest. Here protestors calling for anti-pollution measures picket a motor show.*

● Putting on the pressure

Concerned groups of people have campaigned on environmental issues for many years. One famous international group is Greenpeace and they describe their aims as follows:

> Greenpeace is an environmental pressure group which tries to bring to the attention of the public the way in which the world is being destroyed. We do this by carrying out what we consider to be peaceful direct action. This means we physically interfere. We actually get in the way of processes and operations we think are environmentally unsound. (David Bellamy and Brendan Quayle, *Turning the Tide.*)

Other groups have different ways of promoting their cause – through lobbying politicians, education or practical conservation work. Pressure groups have been very effective in raising public awareness of issues affecting the environment. They have also been effective in mobilizing action against both governments and industries.

International co-operation

There is an increasing acceptance that many problems of pollution control and conservation cannot be resolved by countries acting independently. Transboundary pollution, protection of the seas, loss of habitats and threats to the Earth's ozone layer are problems which have to be met by international effort. (British Department of the Environment, *Protecting Your Environment.*)

The major international environmental organizations

United Nations Environment Programme (UNEP) has its headquarters in Nairobi, Kenya. It co-ordinates and monitors environmental activity throughout the world. It hosts conferences and sponsors international conventions.

The World Conservation Union is based in Switzerland. It is an association of governments, government agencies and environmental organizations. It is involved in international monitoring of the environment, planning and promoting conservation, and providing advice and assistance.

The Organization for Economic Co-operation and Development (OECD) is based in Paris and it brings together European countries, North America, Japan, Australia and New Zealand. It has a specialist Environment Committee and is particularly involved in environmental and economic policies.

The European Community (EC) has had a major programme of environmental protection since 1973. It is now on its Fourth Environmental Action Programme to improve all aspects of the European environment.

> Green power to stop a product dead in its tracks was dramatically demonstrated by the Natural Resources Defence Council (NRDC). Its lobbying helped to bring about a television programme on the alleged carcinogenic [cancer-causing] properties of Alar, a chemical additive to apples. As a result, in 1989 Uniroyal withdrew Alar from use in foods worldwide. Millions of gallons of apple juice were destroyed. (*The Economist*, 20 October, 1990.)

Needless to say, pressure groups such as Greenpeace aren't always popular with those whom they seek to expose. In 1985 the Greenpeace boat *Rainbow Warrior* was blown up by agents of the French Government. Such drastic action was taken to stop Greenpeace protesting about damage to the environment from nuclear weapons testing carried out in the Pacific Ocean. Some pressure groups are also questioned by environmentalists:

> If someone in the green movement says: 'Join us and we will sort out everything on your behalf', they are not doing anyone any favours at all. If instead they say: 'Join us and then work with us to sort out these problems on behalf of the planet and all its inhabitants', that is fine and they have got the message right. It is wrong to encourage passivity [sitting back] and allow people the illusion that merely by paying an annual subscription they have discharged their responsibility to the environment. (Jonathon Porritt, former Director of Friends of the Earth, quoted in *Environment Now*, No. 12.)

• Green voters and green politics

Pressure group action and media coverage have been very successful in informing people about what is happening to the

The President and Vice-President of the USA, George Bush and Dan Quayle, and their wives, at the Republican Convention. Voting for parties with green policies and writing to politicians is one way that people can help bring about changes in environmental laws.

Opposite *The Green Party says, 'Other political parties promise the Moon but only the Green Party guarantees the Earth'. They have been increasingly successful in Europe with MPs in many countries. Their success has also influenced other political parties to consider environmental issues more seriously.*

planet. In recent years there has been a very large shift in public opinion.

> Very large majorities [in a survey] – between 75 and 100 per cent of both the public and leaders in all 14 countries of the European Community – agreed on the need for stronger action by their governments, stronger action by international organizations such as the United Nations and stronger laws to contain industrial pollution.
>
> Almost 97 per cent of Americans surveyed think their country should be doing more 'to protect the environment and curb pollution'. (United Nations Environment Programme, *Our Planet*.)

Politicians have responded to this shift in public opinion. The American President, George Bush, built part of his election campaign around a promise to clean up the USA. Mikhail Gorbachev, the Soviet leader, made the environment a key theme of an address to the United Nations. In 1990 the British Government published *This Common Heritage* as an outline for a programme of action to deal with environmental problems.

Daily Mail

MONDAY, NOVEMBER 27, 1989

25p

Read Keith Waterhouse PAGE EIGHT

THE GREATEST CAR CONTEST IN HISTORY (if you're 12 or under)

INSIDE TODAY

WIN THE ULTIMATE JUNIOR DREAM CAR

PLUS FIVE MORE MOTOR-DRIVEN CLASSICS TO BE WON

TURN TO PAGE 32

Hypermarkets threatened in bid to help environment

PATTEN'S WAR ON POLLUTION

By GORDON GREIG and MICHAEL KEMP

A MAJOR offensive against pollution, giving Britain the most forward-looking 'green' policies in Europe, was outlined yesterday by Environment Secretary Chris Patten.

But one element could be highly controversial — the curbing of out-of-town shopping at hypermarkets.

Planning permission for stores could be severely restricted as part of a campaign to cut unnecessary motoring.

But Mr Patten would be going against one of the major shopping trends of the past 20 years. The stores and their giant car parks have spread spectacularly and are hugely popular. Many housewives, particularly those with freezers, are prepared for a lengthy drive so that they can buy cheaply in bulk and would not want to return to the days of struggling on public transport laden with shopping bags.

Attack

Last night major chains planning out-of-town stores, like Marks & Spencer, Sainsbury, and W. H. Smith, said they were concerned at Mr Patten's statement and called for a clear statement of Government policy on the issue.

Mr Patten is determined to check the unbridled growth of motoring. He said on TV yesterday: 'Do I go on approving out-of-town shopping centres when one knows they are going to have a considerable effect on vehicle mileage?'

He is now considering a new pollution tax based on engine size to encourage the buying of smaller cars. Petrol tax will go up by 7p a gallon in the Budget next March and continue rising through the 1990s. Next

Turn to Page 2, Col 6

GREENPRINT FOR THE FUTURE

These are some of the ideas being studied by the Government which could appear in a discussion paper early next year.

- Changing car-dominated shopping patterns by vetoing plans for new superstores on 'green field' sites.
- Tax perks for companies adopting 'green' anti-pollution measures.
- Toll payments for the most heavily-used city roads.
- Dearer road fund licences with the tax tied to engine size. The present £100 for cars could go up on a sliding scale to £200-£250 for 'gas guzzlers'.
- Europe-wide rules curbing the toxic exhaust gases on all new cars.
- Heavier duties on fuels such as gas and electricity to promote energy conservation and avoid having to build more power stations.

Batman star Kim Basinger: Seen out with other men

Actress Kim 'in £15m divorce'

From GEORGE GORDON in New York

ACTRESS Kim Basinger, star of Batman and 9½ Weeks, is facing a bitter divorce battle that could cost her half her estimated £30million fortune.

Miss Basinger and actor-producer Ron Britton have been living apart for several months after seven years of marriage.

Recently the 35-year-old star has been seen with wealthy Batman producer Jon Peters, pop singer Prince and Australian fitness coach Phil Walsh.

Britton, 50, is reportedly demanding a multi-million dollar settlement. Under Californian law he is entitled to half the property and money acquired during the marriage.

If he doesn't get an out of court settlement, he is prepared to go to trial and present a list of men allegedly involved with his wife.

Miss Basinger's biggest headache in the settlement will be the £14million she spent on buying a whole town, Braselton, in Georgia, where she grew up.

One of Britton's friends said: 'Ron is bitter and wants everything he is entitled to.'

INSIDE: Weather 2, World Wide 10, Femail 13, Diary 23, TV and Radio 26-27, Letters 34, Headlines Game 36, City 39, Sport 40-48

Left *This headline highlights the 1989 launch of the British Government's policy on the environment, put forward by the then Environment Secretary Chris Patten. Do you think ordinary people are prepared to make changes in their lifestyles, or will they only change if they are forced to by law?*

Many political parties now have 'green manifestos' and there is a successful Green Party movement throughout the world.

> Now there are Green MPs [Party representatives] sitting in twelve national parliaments and, with the exception of Greece, there are Greens elected to some level of local government in every West European country. (Sara Parkin, speaker for the British Green Party, quoted in *Environment Now* No. 12.)

Now that public awareness of environmental issues is higher, voters are more likely to choose a political candidate with a good environmental policy. But is the concern shown by

Some 14 per cent of Americans now play an active part in green organizations. The American National Wildlife Federation (NWF) is the world's biggest environmental group, which had 4 million supporters in 1980 and 6 million in 1990. Its annual budget in 1990 was $100 million.

In Britain, Greenpeace has about 300,000 members and Friends of the Earth 200,000. More people belong to environmental groups than to political parties.

politicians genuine or are they just jumping on the bandwagon? Are they merely looking for voters to put them into power and keep them there? And are the voters quite as green as they'd like to think they are?

> The political facts of life are stark. Most voters are all for 'green' values so long as they make no inroads on [do not affect] their material standards of living . . . (Patrick Rivers, *The Stolen Future*.)

The influence of the green movement has been widespread. Change is happening. There are now a host of environmental laws and conventions and pollution control regulations. Waste reduction and recycling schemes are very popular, as are environmental education programmes and conservation projects. But are these enough?

> Ultimately, responsibility for the future rests with individuals. Our values, choices, and behaviours shape social and political change. Unless more of us join the effort, there is little hope of halting the planet's deterioration . . .
> . . . The ultimate rationale for a massive social mobilization to safeguard the earth is summed up in a bit of graffiti painted on a bridge in Rock Creek Park in Washington, D.C. It says, 'Good planets are hard to find'. (Lester R Brown *et al, State of the World 1989*, Worldwatch Institute.)

International solutions are needed to many environmental issues. Dr Mostapha Tolba, the Director of the United Nations Environment Programme (UNEP), addresses an international conference on atmospheric pollution and the environment in 1989. Environment ministers from more than sixty countries attended.

A chemical factory producing medicines alongside a river discharges its waste into the water. The pollution is killing fish, and local fishermen find that they can no longer earn a living. You are the Minister of the Environment. How would you reply to the fishermen's letter of protest?

6

Cleaning up

There are two broad approaches to controlling pollution and tackling environmental problems. Firstly, national and international laws can be passed to establish environmental standards that must be kept. Secondly, taxes and prices can be used as a market influence on producers and consumers.

- **Environmental laws**

Laws to limit pollution have been the foundation of control measures at national and international levels. The European Community has initiated 280 measures to protect the environment since 1973; and on a global scale there are international agreements such as the Montreal Protocol, which commits signing nations to reduce their output and use of chlorofluorocarbon (CFC) gases, which damage the ozone layer. Pollution control laws force industries to clean up. Many companies, especially multinationals, have responded.

> Because of the environmental pressures on hundreds of its products, from pesticides to aerosol propellants . . . the greening of the chemical industry has been proceeding at a spectacular pace. (John Elkington, *The Green Capitalists*.)

Pollution control is expensive but not as expensive as cleaning up after the event. Prevention is better than cure.

> One of the key justifications for environmental action is the cost of doing nothing, the price of waiting for the pollution bill to come thudding on the doormat. (*Environmental Policy in the European Community*, European Documentation, May, 1990.)

Cleaning up after a pollution accident can be very costly. It is estimated that the bill for cleaning the Alaskan coastline after the Exxon Valdez *oil spill cost Exxon about US$1.3 billion. Preventing pollution happening in the first place is a cheaper option.*

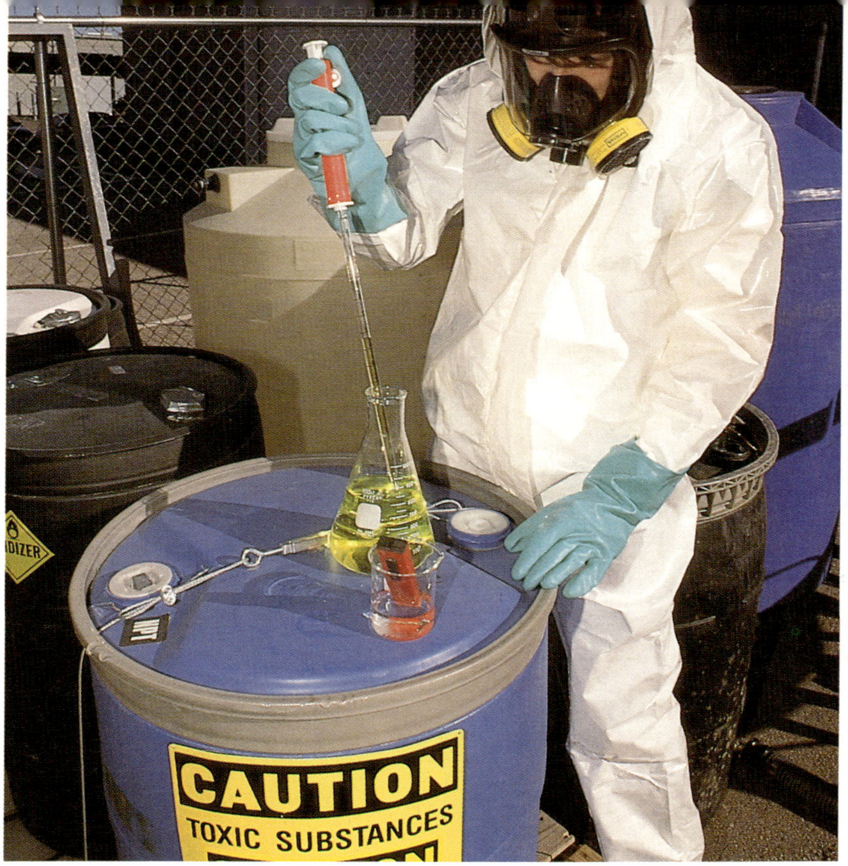

Many companies behave responsibly when it comes to pollution control. They allocate large amounts of their budget and workforce to environmental protection. Here, toxic waste chemicals are being tested to assess their affect on the environment.

Many companies have now come to realize that it can pay to be clean. Large companies represented in many countries are now spending staggering amounts on pollution control. For example, Bayer, the German chemicals group, spends 20 per cent of its manufacturing costs on measures to protect the environment. The largest chemicals group in the world, ICI, has been anxious to develop a greener corporate image and show that it cares for the environment. It has given top executives strict environmental targets to meet.

Multinationals may be able to bear the high cost of pollution control measures but what about smaller companies? They may not be able to afford to invest in cleaner technology or to produce less environmentally harmful products. If the cost of cleaning up is too great, the company will go out of business and jobs will be lost. Cleaning up the environment is certainly not a simple, cheap process.

What price are we prepared to pay for pollution control? Is it acceptable that jobs are lost and businesses close? And who should pay – the industries, the government or the consumer through taxation?

The business of pollution control

In the European Community, between 1.5 and 2 million people are employed in pollution control and clean technology industries. In the USA in 1985, 167,000 people were employed in air, water and solid-waste pollution control. This created US$19 billion in sales and US$2.6 billion in company profits.

(Source: Ambio)

• The Polluter Pays Principle

One policy that can be used to control pollution is the Polluter Pays Principle. This principle states that:

> The cost of preventing pollution or of mitigating [lessening] environmental damage due to pollution should be borne by those responsible for the pollution. (*National Society for Clean Air Reference Book.*)

But some environmentalists disagree:

> . . . there are serious limitations to this [the Polluter Pays] approach. Many forms of pollution are a combination of substances from diverse sources. Moreover, the contribution from an individual enterprise or household is often negligible; the damage arises out of cumulative additions. It would be impossible to calculate fairly who should pay what costs. And, critically, some consequences of pollution are irrevocable, and no amount of taxes can make amends even if we could put a price on what has been lost. (Irvine & Ponton, *Green Manifesto.*)

The Rechem incinerator plant at Fawley in England disposes of toxic chemical waste by burning it at very high temperatures. Britain imports toxic waste from other countries. Do you think every country should dispose of its own waste or is exporting it acceptable?

Old refrigerators containing polluting CFC gases are stored before the gas is extracted to be recycled. The release of CFCs into the air upsets the balance of ozone in the upper atmosphere. Recycling CFCs means less new gas needs to be manufactured.

• Pollution trading and clean technologies

In many countries, laws and regulations set out emission standards. These are the maximum permitted levels of pollution allowed. Another pollution control measure is pollution trading:

> Pollution trading allows some factories to pollute above emission standards [the permitted amount] so long as others are below these standards. Some pollution trading may be desirable but most will not since practice involves 'trading' in damage to humans or ecosystems between different localities. (Friends of the Earth, *Earth Matters*, Autumn 1989.)

The development of cleaner technologies could also reduce pollution:

> The solution to this [pollution] situation requires the development of technological processes that do not create environmental contamination . . . (*Canada's Environment: An Overview.*)

Waste reduction schemes and an increase in recycling also help to combat pollution:

> The environmental pressure is forcing growing numbers of companies to look at ways in which they can design toxic waste processes. Dow Chemical, for example, produced one pound [0.454 kg] of waste for every pound of product in 1960. By the 1980s, however, it was producing one pound of waste for every 1,000 pounds [454 kg] of product. (John Elkington, *The Green Capitalists.*)

Increasingly, cleaning up is seen by industry as good business:

> Good environmental management is an investment. It can improve your company's reputation, boost employee morale and attract good quality recruits, increase efficiency, or cut costs by eliminating avoidable waste. It can even improve your profits. (Confederation of British Industry, *Clean Up – It's Good Business.*)

Which of these statements do you agree with?

a) I would be willing to pay 10 per cent more for goods produced by clean technologies that do less damage to the environment. I am prepared to be 10 per cent poorer.

b) I am willing to stay as I am and neither increase nor decrease my wealth and standard of living.

c) I want to improve my standard of living. I am only prepared to take steps to protect the environment after I have achieved the standard of living I want.

7 The greenhouse effect – a case study

> The year is 2035. World food supplies are at an all-time low as wheat production in North America is decimated by climatic change . . . The population of the United Kingdom basks in sub-tropical heat – and queues for water rations . . . One and a half million people died in the floods that devastated Bangladesh in 2032. The survivors have joined the millions of environmental refugees roaming the planet. The Environmental Security Council of the United Nations was formed in 2015 in a desperate bid to curb global warming. It has still failed to reach agreement on the measures needed. (Jeremy Rifkin & Mick Kelly, *The DIY Guide to Combating Global Warming.*)

Science fiction or a realistic view of life in the twenty-first century? It depends on your point of view. The greenhouse effect, or global warming, has been called the greatest challenge to face humankind. It has become the major environmental issue of this century because it threatens to upset the world climate.

In this chapter we will use the greenhouse effect as a case study to see how the issues of morality, economics, politics and development relate to the central problem – pollution.

Some scientists predict that global warming will cause parts of the polar ice sheets to melt, raising sea levels throughout the world.

A coking plant in Wales, UK. Our society consumes large amounts of fuel, but burning fossil fuels increases CO_2 levels in the atmosphere. CO_2 has increased from 265 parts per million (of air) in 1850, to 350 parts per million in 1988. It is estimated that this extra CO_2 causes 49 per cent of the 'greenhouse' problem.

• What is the greenhouse effect?

Scientists measuring the global climate have found that the world's climate is changing. The average world temperature has increased by about half a degree Celsius since the 1850s. Scientists are predicting that this trend will continue. By the middle of the next century it is likely the world will be 1.5 °C warmer than today. Not a big change, you might think; however, in the 10,000 years since the last ice age, the Earth has only warmed up by 4 °C. How is this warming caused? The quotation below explains:

> Polluting gases, released into the atmosphere as a result of human industrial activities . . . trap heat near the surface of the planet, and prevent heat from escaping into space. The process is known as the greenhouse effect, and the gases responsible for this heat trap are known as greenhouse gases. Since greenhouse gases are being released at ever-increasing rates today, the warming of the world experienced during the twentieth century is likely to be only the beginning of a much bigger, and much more rapid, shift in the world's climate. (John Gribbin, *The Global Greenhouse*.)

There are about thirty greenhouse gases, but the main one is carbon dioxide; others include methane, nitrogen and chlorofluorocarbons (CFCs). Where do they come from?

> Man, by burning fossil fuels and rainforests, has increased the amount of carbon dioxide in the atmosphere by more than 25 per cent since the Industrial Revolution. Man, by growing ever more rice, by raising ever more cattle, has more than doubled the amount of methane in the atmosphere. So, in essence, we have a new atmosphere. We do not know how this new atmosphere works. (Bill McKibben, *The Independent*, 27 January, 1990.)

• The need for evidence

The need for scientific evidence of global warming has delayed action. Gathering and analysing evidence takes many years because the Earth's climate is very complex. Not so long ago, some scientists were predicting that the Earth was cooling down and heading for another ice age! Now there are signs that the opposite is true. But, however much research is needed, should we delay acting to save the environment?

> The fact that scientists and decision-makers are currently facing knowledge gaps should not deter countries from taking immediate action . . . (Dr Mostafa Tolba, editorial in *Our Planet*, Vol. 1, No. 4, 1989.)

Most of the world's scientists now agree that the greenhouse effect is happening but there is still some disagreement:

> A handful of . . . scientists, mainly in the United States, dispute the consensus and have been taken up by the media eager to challenge orthodoxies [conventional opinion], and by energy industries anxious to preserve the growth of their business. (Geoffrey Lean, *Country Life*, 22 November, 1990.)

Global research is under way. The United Nations Environment Programme has set up the Global Environment Monitoring System (GEMS) which gathers and distributes environmental data worldwide. There are also many other research programmes looking at climatic change and its effects.

Rice terraces in Bali. The rotting plant and animal material from rice paddy fields gives off methane, a greenhouse gas. As rice production has grown worldwide, so has the amount of methane in the atmosphere.

Industrialized nations are responsible for 90 per cent of greenhouse gas emissions although these countries make up only 23 per cent of the world's population.

> A few countries have so far accounted disproportionately for the rise in carbon dioxide but . . . the resulting climatic changes will not be allocated to any earthly idea of justice. (Erick Eckholm, *Down to Earth*.)

Some of the worst effects may be felt in developing countries. As the world becomes warmer, the seas will expand and the polar ice caps are likely to melt, causing sea levels to rise. Tuvalu, a small group of low-lying islands in the Pacific Ocean, will be covered over by sea.

> Tuvalu boasts only fifteen of the world's 400 million cars yet it is the burning of oil and other fossil fuels that is the main cause of the greenhouse effect. The country's fate is being sealed by the casual daily activities of hundreds of millions of people, thousands of miles away, most of whom do not . . . know of its existence . . . it can do nothing to save itself . . . (Geoffrey Lean, *Country Life*, 22 November, 1990.)

The Prime Minister of Tuvalu, Bikenibeu Paeniu, says:

> We contribute nothing to this problem, and yet we will be the first to suffer.

Tuvalu is not alone. As the greenhouse effect increases, other island groups are predicted to disappear and flooding will drown many coastlines around the world. What responsibility does the developed world have towards other people on the planet? How would you feel if you lived on Tuvalu?

Scientists predict that global warming will affect the weather, vegetation and agricultural patterns around the world. Weather conditions could become more extreme with hurricanes (seen here) and heavy winds, or with droughts.

• Greenhouse solutions

The only solution is to cut down the amount of greenhouse gases we discharge into the atmosphere. This seems obvious but why is it so difficult? One reason is that the costs are enormous.

> The US has promoted the view that an immediate policy response to the threat of global warming could bankrupt Western economies and cost trillions of dollars. (Greenpeace, *Campaign Report*, November, 1990.)

To date, there have been two World Climate Conferences, where world governments have met to attempt to reduce the emission of greenhouse gases worldwide. There has been little real progress:

> While most nations are willing to accept the inevitable consequences of global warming if carbon emissions are not reduced immediately, some countries find it difficult to make critical commitments vital to solving the crisis. (Tessa Robertson, *WWF News*.)

Flooding is predicted as another effect of global warming. As the planet heats up, the seas will expand and ice sheets will begin to melt, raising sea levels and flooding low-lying areas. This would cause widespread environmental damage and loss of life.

An experimental wind farm near Urumqi, Xinjiang Province, in the north-west of China. Reducing CO_2 levels in the atmosphere is the solution to the greenhouse effect. This may mean using less energy from fossil fuels. Energy conservation programmes could achieve this, as could the development of alternative energy sources such as wind power.

> I think proposals to limit CO_2 [carbon dioxide] emissions at present levels to the year 2000 and to reduce them by 20 per cent by 2005 are too hasty! There are so many things we do not understand about the relationship between CO_2 and the greenhouse effect. (Okamatsu Sozaburo, Japanese Ministry of International Trade and Industry, quoted in *Our Planet* (UNEP).)

When the course of action needed seems so obvious, why have so few governments responded?

> . . . to slow global warming in the next century, big, uncomfortable decisions need to be taken now . . . The trouble is that the difficult decisions could make today's politicians deeply unpopular with voters, as well as with industry and commerce. (Nicholas Schoon, *The Independent*, 17 April, 1990.)

Power, politics and money are all ingredients in the global warming dilemma but it is easy to blame governments and politicians. The solution also lies in our own hands. Are you prepared to travel by car less often, to use less electricity, not to travel abroad on aeroplanes, to consume less?

> What can be stated is that the cause of the greenhouse effect is not just coal or cars or cutting down forests but a whole range of man's activities. The solutions will therefore have to be global. Everyone can help. Each individual effort will of course seem insignificant, but when multiplied millions of times the effect is considerable. (Francis Wilson, *Clouds; the Key for the Future*, published in *The Field*, December, 1990.)

Which do you think is the most important global issue facing you today and why?
a) unemployment
b) nuclear war
c) the greenhouse effect
d) racism
e) destruction of the rainforests
f) homelessness
g) world poverty
h) terrorism

8 The future of the planet

> At the moment we are afraid. Do not leave our generation without hope. Have the courage to put aside your short-term national concerns and act in the interests of our common global well-being. Will you protect our future? (Susannah Begg of the Australian Youth Delegation to the Bergen Conference, May 1990.)

This cry from the heart came at a regional conference organized by the World Commission on Environment and Development. It reflects how many young, and older, people feel about the state of the world environment. We are bombarded with doom and gloom about the planet. How serious is the pollution threat? Is the future of the planet really threatened? This is the view put forward in 1990 at an American conference 'Preserving the Global Environment':

> If the world community fails to act forcefully in the current decade, the Earth's ability to sustain life is at risk. (The American Assembly/World Resources Institute.)

How serious is the pollution threat to the Earth? Have we reached the point of no return, or can concerted individual and international action save the planet?

Antarctica is a potent symbol of the environmental crisis. It has been called the last great wilderness on Earth. Many conservationists want this unique area protected forever as a world park, but many governments want to begin mining operations there. Its future is uncertain.

David Attenborough, the British naturalist says:

> In our hands now lies not only our own future, but that of all other living creatures with whom we share the Earth. (David Attenborough, *Life on Earth*.)

Others feel that, although humans can do themselves and much of the planet immense damage, they could not extinguish all life on Earth.

> . . . if we seriously foul our nest we shall not destroy life on Earth – other organisms will flourish in their environments – but we will destroy ourselves. (James Lovelock, quoted in *The Independent*, 23 September, 1989.)

• The Gaia theory

Professor James Lovelock is a well-known independent scientist. He has put forward a different view of life on Earth, which he calls Gaia. For Lovelock, the Earth is to be respected as a living being. It grows and develops, supporting itself and all other forms of life, which depend on it for survival. Humans may like to think they are the controllers of the environment, the most important species on Earth, but in the Gaia theory they are not:

> If we think only of ourselves and degrade the Earth, then it will respond by replacing humans with a new and more amenable species. It will live, but we may die. (James Lovelock, quoted in *The Independent*, 23 September, 1989.)

A positive response to the environment. The National Centre for Alternative Technology in Machynlleth, Wales, tests and demonstrates alternative technology, such as this solar-powered cottage.

Gaia is a controversial theory and many scientists disagree. However, the following quotation illustrates the new thinking that Gaia has encouraged:

> If anything, the concept of Gaia serves to remind the human race that though the environment is in many ways fragile . . . she can respond to our inappropriate actions in deadly self-defence. It is the health of the planet that matters to Gaia, and not the health of any particular species that inhabits it! (Anne-Maria Brennan, *Environment Now*, March 1990.)

• Good management

Whatever you think about the long-term future of the planet, it makes sense for humans to act now to improve their local and global environment. Environmental damage is caused by ordinary people doing ordinary things – the problem is that there are millions of people doing them! The situation is serious but not impossible. Here are some people's idea of what action is needed. Chris Patten, Britain's former Secretary of State for the Environment, sees the need for informed action:

> Effective concern for the environment requires deep commitment, cool judgement and the sound application of scientific principles. If we proceed by panic and scaremongering, we will only serve to discredit the cause of the environmental movement. (Chris Patten, quoted in *The Independent*, 15 August, 1989.)

Dr Bankowski calls for new ways of thinking:

> Perhaps the worst environmental pollution is pollution of the mind, and the greatest need is for well thought-out principles of environmental ethics . . . to provide a basis for the protection and preservation of nature, rather than its exploitation for what humans consider to be their needs and interests. (Dr Zbigniew Bankowski, *World Health*, January/February, 1990.)

The Dalai Lama reminds us all of our responsibilities:

> When we talk of a global crisis, we cannot blame a few politicians, a few fanatics or a few trouble-makers. The whole of humanity has a responsibility because this is the business of us all. (His Holiness the Dalai Lama.)

Is the environment everybody's responsibility or are we, as individuals, powerless against the interests of governments and multinational companies?

> Hopelessness and helplessness are predictable responses to the current state of affairs. What can one individual do to alleviate the symptoms of the massive environmental decline of the last 40 years? (*Friends of the Earth Handbook*.)

Good management of the environment requires looking at reusing, recycling, repairing and reclaiming resources, rather than using new ones and throwing them away. New technologies, such as this waste oil reclamation plant, have to be developed to do this.

● What can one person do?

> . . . many of us are beginning to recognize that every action we take can affect the well-being of the planet. Some things we can do on our own whereas others require groups of like-minded people. Still others need pressure to be exerted on those in power in our society. But there is no end of things we can do. And of course they don't all have to be done at once. You can begin in small ways, taking it one step at a time. (John Button, *How to be Green.*)

There are many things each of us can do, from writing to our politicians to cycling instead of driving a car. Action is the key:

> Action means different things to different people. One person might stop buying a particular product . . . Another might start a discussion group at school or college. Another might choose one particular subject and work on a campaign . . . By taking action, you will be making it clear to people what you think and believe – and you will be helping the planet too. (Debbie Silver and Bernadette Vallely, *The Young Person's Guide to Saving the Planet.*)

Do you always need to go by car? Using public transport is one way of reducing your own impact on the environment. Buses and trains are more energy efficient than cars, cause less pollution and reduce traffic jams.

Save it. We can all help save resources and reduce pollution by recycling cans, bottles and paper.

• What YOU can do to be less polluting

Conserve energy

- Switch off heaters and lights when they are not being used.
- Wear an extra sweater rather than turn up the heating.
- Use public transport, walk or cycle when you can, rather than go by car.
- Take a shower rather than a bath – showers use less water and therefore less heating.

Conserve resources

- Do not buy over-packaged goods. Refuse extra paper or plastic bags in shops.
- Take used bottles, cans and paper to a recycling centre.
- Cut down on the amount of disposable goods you buy – reuse goods whenever possible.
- Buy recycled paper and paper products.

Find out more

- Learn more about the environment and pollution problems by keeping up with the news. There are plenty of good books to read. Some suggestions are on page 47.
- Talk about environmental issues with your friends and family. You can learn a lot this way.
- Find out about green organizations and conservation groups. You may like to join a group you particularly support.

What do you think YOU can do to protect the well-being of the environment and its inhabitants? Why not make a list of ideas and check up on it regularly, to see how you are meeting your challenges?

Glossary

Acid rain Rain, snow, mist or hail which is made more acid by waste gases.
Biodegradable The term given to a substance which can be broken down by decomposition.
Biosphere The part of the Earth and the atmosphere in which plants and animals live.
Carbon dioxide A gas found in the atmosphere; it can be formed from burning fossil fuels.
Cash crop A crop grown for sale abroad rather than for food in the area where it is grown.
Chlorofluorocarbons (CFCs) The group of chemicals that damage the ozone layer.
Conservation The protection and careful use of resources, so as not to damage the environment.
Ecosystem A community of plants and animals and the environment they live in.
Emissions Substances discharged into the air from factories, domestic chimneys and car exhausts.
Fossil fuels Fuels derived from organic (once living) material such as oil, coal and natural gas.
Global warming The increase in the average temperature of the Earth thought to be caused by the build-up of greenhouse gases in the atmosphere.
Greenhouse effect The warming of the Earth caused by the accumulation of certain gases in the atmosphere.
Greenhouse gases The name for about thirty atmospheric gases which allow the heat of the Sun to reach the Earth but trap the heat escaping back out to space.
Nitrogen oxides (NOx) Gases formed mainly from nitrogen in the atmosphere when fuels are burnt at a high temperature.
Ozone (O_3) A form of oxygen that can damage plant growth and irritate the eyes and breathing system. However, ozone forms a vital layer of the atmosphere, shielding the Earth from the Sun's harmful ultraviolet rays.
Sulphur dioxide (SO_2) A colourless gas with a strong odour. It is mainly formed from burning fossil fuels.
Ultraviolet radiation (UV) Harmful but invisible light waves coming from the Sun. Most are screened out from the Earth by the ozone layer.

Further information

You can contact these organizations to find out more about pollution and the environment.

Australia

Friends of the Earth, 366 Smith Street, Collingwood, Victoria 3066
Wilderness Society, 130 Davey Street, Hobart TAS 7000

Canada

Ecology Action Center, 1657 Barrington Street, Suite 520, Halifax, Nova Scotia B3J 2A1
Friends of the Earth, 251 Laurier Avenue West, Suite 701, Ottawa, Ontario K1P 5J6
Greenpeace, 427 Bloor Street West, Toronto, Ontario M5S 1X7

New Zealand

Friends of the Earth, PO Box 39/065, Auckland West
Greenpeace, Private Bag, Wellesley Street, Auckland
World Wide Fund for Nature, Box 6237, Wellington, New Zealand

United Kingdom

British Trust for Conservation Volunteers, 36 St Mary's Street, Wallingford, Oxfordshire OX10 0EU
Council for Environmental Education, University of Reading, London Road, Reading RG1 5AQ
Friends of the Earth 26-28 Underwood Street, London N1 7JQ
Greenpeace 30-31 Islington Green, London N1 8XE
National Society for Clean Air, 136 North Street, Brighton BN1 1RG
Tidy Britain Group, The Pier, Wigan WN3 4EX
World Wide Fund for Nature, Panda House, Weyside Park, Godalming, Surrey GU7 1XR

USA

Environmental Defense Fund, Suite 150, 1616 P Street NW, Washington, DC 20036
Greenpeace, 1611 Connecticut Avenue NW, Washington, DC 20009
Sierra Club, 330 Pennsylvania Avenue NW, Washington, DC 20005

Further reading

Allaby, Michael *Green Facts* (Hamlyn, 1989)
Bellamy, David and Quayle, Brendan *Turning the Tide* (Collins, 1986)
Bright, Michael *The Dying Sea* (Franklin Watts, 1988)
Button, John *How to be Green* (Century, 1989)
Conserving Our World series (Wayland, 1989)
Considering Conservation series (Dryad Press, 1987)
Elkington, John and Hailes, Julia *The Green Consumer Guide* (Gollancz, 1988)
Gribbin, John and Kelly, Mick *Winds of Change* (World Wide Fund for Nature, 1989)
Hawkes, Nigel *Toxic Waste and Recycling* (Franklin Watts, 1988)
McCormick, John *Acid Rain* (Franklin Watts, 1986)
Seager, Joni (ed) *The State of the Earth* (Pluto Press, 1990)
Seymour, J and Giradet, H *Blueprint for a Green Planet* (Dorling Kindersley, 1988)
Silver, Debbie and Vallely, Bernadette *The Young Person's Guide to Saving the Planet* (Virago, 1989)

Teaching Materials

Tilling, Stephen *Ozone and the Greenhouse Effect: a practical GCSE coursework guide* (Field Studies Council, Preston Montford, Montford Bridge, Shrewsbury SY4 1HW, 1990)
Air Pollution Pack and *Noise Pollution Pack* (Available from the National Society for Clean Air, 136 North Street, Brighton BN1 1RG, 1987)
Environment wallcharts: a series of titles on pollution, the ozone layer and greenhouse effect are available from Pictorial Charts, 27 Kirchen Road, London W13 0UD
Audio-visual packs: a series of titles on environmental issues is available from the International Centre for Conservation Education, Greenfield House, Guiting Power, Cheltenham GL54 5TZ
Resource sheets: a series of pollution titles is available from the Council for Environmental Education, University of Reading, London Road, Reading RG1 5AQ

Picture acknowledgements

The publishers would like to thank the following for providing the illustrations in this book: John Baines 39; Bruce Coleman 6 (Joy Langsbury), 8, 15 (Michael Freeman), 17 (Michael Freeman), 18 (Luiz Claudio Marigo), 21 (Robert P. Carr), 33 (Thomas Buchholz), 34 (Francisco J. Erize); John Frost 26, 28; Geoscience Features 11, 35, 38; Greenpeace 12, 24, 30, 41, 45; Rex Features 4 (Greg Mathieson); Topham 5, 9, 13, 14, 27, 42; Wayland Picture Library 16, 19, 20, 36, 39; WHO 7, 23, 25; ZEFA *cover*, 10, 37, 40, 43, 44.

Index

Page numbers in **bold** refer to illustrations.